Al Jolson, the popular American singer of the 1920s to 1940s, features in this working model made around 1948 and called 'Jolson Sings'. On insertion of a coin, the puppets appear to sing and play the banjo while a recording of a famous Jolson song is heard. The suddenness of the initial movement is always startling for the observer.

AMUSEMENT MACHINES

Lynn F. Pearson

Shire Publications Ltd

CONTENTS

Printed in Great Britain by C. I. Thomas & Sons (Haverfordwest) Ltd, Press Buildings, Merlins Bridge, Haverfordwest, Dyfed SA61 1XF.

British Library Cataloguing in Publication Data: Pearson, Lynn F. Amusement Machines. — (Shire Albums Series; No. 285) I. Title. II. Series. 688.7. ISBN 0-7478-0179-7.

ACKNOWLEDGEMENTS
The author wishes to thank Jon Gresham of Penny Arcadia for his invaluable comments throughout the writing of this book, and John Hayward of the National Museum of Penny Slot Machines for his generosity with information and illustrative material. The author also wishes to thank the British Amusement Catering Trades Association, Great Yarmouth Central Library, Helene Hogg, Scarborough Borough Council, Sotheby's and York Castle Museum for their assistance.
   For permission to reproduce photographs of amusement machines in their collections, the author wishes to thank Bryan's Works of Kegworth Museum, Drayton Manor Park Ltd, pages 4 (bottom), 5, 11 (top), 18 (all), 19 (top), 22 (all), 23 (left) and 24 (both); Chewton Cheese Dairy, page 11 (bottom right); Eastbourne Heritage Centre, pages 9 (left), 13 (bottom right), 17 (both) and 20; Michael Ford, pages 28 (all), 30 (all) and 31 (both); Jon Gresham of Penny Arcadia, pages 8 (bottom), 9 (right), 10 (right), 12 (top two), 14 (top right), 16 (top) and 26 (both); Carol Johnson of Reds Rock Cafe Bar in Newcastle upon Tyne, page 19 (bottom two); National Museum of Penny Slot Machines, John Hayward Collection, pages 8 (top), 11 (top), 13 (bottom left), 14 (top left), 18 (all), 19 (top), 22 (top), 24 (both) and 29; the Comptroller of the Patent Office, pages 12 (bottom) and 21 (both); Woodspring Museum, pages 2 and 4 (top); Wookey Hole Caves Ltd, pages 10 (left), 14 (bottom two), 16 (bottom two), 23 (right), 25 and 27 (both); and York Castle Museum, pages 1, 7 and 11 (bottom left). The cover photograph is by the author.

Cover: *An Art Deco styled fruit machine by O. D. Jennings & Company in use at the Town Moor Hoppings, a fair in Newcastle upon Tyne, in June 1991. The first version of this machine was produced by the American firm in 1949, and variations on the same theme were manufactured until the late 1960s.*

Left: *The interior of the Regent Street amusement arcade, Weston-super-Mare, in the 1920s. In the centre is a fortune-teller, the Human Analyst, and in the left foreground an early form of juke-box. Almost all the machines are freestanding rather than fixed to the arcade walls as they would increasingly be from the 1930s.*

Among the attractions on Brighton Palace Pier (opened in 1899) were the amusement machines strategically sited along its length. In this Edwardian view machines are visible in the foreground on the far right and far left; both are freestanding machines with pedestal supports. Later machines were sited in pavilions, away from the elements, and often attached to the walls of these forerunners of the arcades.

ARCADES AND ENTREPRENEURS

Although the seaside penny arcade is usually thought of as the home of the slot machine, amusement arcades originated with the activities of travelling showmen at fairgrounds and in the cities during the late 1880s, when coin-operated amusement machines were first incorporated into their sideshows. The machines were novelties in their own right, as likely to draw a crowd as the more traditional entertainments of freak shows, shooting galleries, waxworks and circus acts. The showmen travelled to country fairs in summer and moved to the cities during the winter, where untenanted shops, hired for low rents, were often used as booths. These shops were the forerunners of the amusement arcades which appeared at the seaside resorts in the 1890s.

Amusement machines were often placed at the entrances of showmen's booths or shops to entice customers inside to the main attraction, perhaps a shooting gallery. By the mid 1890s shops or arcades devoted entirely to amusement machines were popular in London. These frontless shops varied in their degree of decoration, some ornate, others bare, but entry was free, and the public proved keen to sample the increasing variety of automatic amusements, from mutoscopes and model football games to shooting ranges and strength-testers. Initially these automatic vaudevilles were almost silent apart from the noise of the machines, but coin-operated music devices were soon introduced, enabling a music-hall atmosphere to be reproduced at the customer's own expense.

One of the first seaside amusement arcades was the Paradium on Marine Parade, Great Yarmouth, opened in 1902 by George Barron, owner of a firm of amusement-machine manufacturers, Inter-changeable Automatic Machine Company of Islington, London. Barron (1853-1944), the son of a Lincolnshire farmer, married a Yarmouth palm-reader from the Gray family of Norfolk showpeople. He established several arcades in London before his first Great Yarmouth venture, the Jubilee Exhibition, which opened in 1897. The Exhibition was a miniature fairground, with a steam-driven mechanical organ, a giant doorman, a jungle rifle range, 'Beautiful Marie the Giant School Girl', mechanical models, a fancy

3

Above: *Outside the Regent Street amusement arcade, Weston-super-Mare. The arcade, emphasising free admission, was well placed to divert trippers heading from railway station to seafront. A weighing or vending machine stands in front of the central rifle range.*

Left: *The Clock, first produced in 1930 by Bryan's Automatic Works. The player pressed a penny into the disc-shaped hole under the machine name and then turned the handle to free the clock mechanism. The hands on the clock face rotated, and if the minute hand stopped exactly on any of the hours a prize would automatically be paid out through the slot near the bottom of the machine. Stopping at twelve o'clock won the largest prize.*

bazaar and the town's first cinema show.

The Exhibition was destroyed by fire on 5th September 1901 and Barron replaced it the following year with the Paradium, dispensing with the live entertainment. Amusement machines could be found at seaside resorts from the early 1890s; the Scarborough Promenade Pier Company registered its intention to provide machines for its customers as early as 1890. The machines, usually hired on a basis of shared profits, were often situated beside pier entrance buildings or along the sides of the piers themselves. They were mainly freestanding; the later pier or sea-front amusement arcades were ideal situations for wall machines, which required a rear support and became popular after the turn of the century.

Barron's Inter-changeable Company was one of the earliest British amusement-machine manufacturers, though many other firms entered the field in the 1900s and 1920s. Designs patented by G. Haydon and Frank H. Urry were produced by their Automatic Machine Company of London during the 1890s and early 1900s, while Frederick and Arthur Bolland's Amusement Machine Supply Company of London began trading soon after the First World War. By the 1930s the Bollands were advertising over two hundred different machines.

Other major British manufacturers included Charles Featherstone's British Manufacturing Company, a leading maker of wall machines during the 1930s, and Bryan's Automatic Works of Kegworth near Derby, established by William Edward Bryan (died 1984) in 1920. W. E. Bryan was a prolific inventor who produced 49 different amusement-machine designs and patented many improvements to their mechanisms. Beginning with the Oddclod (1927), he went on to design 24 machines in the 1930s and eleven in the 1950s; his last work was Elevenses (1955), an eleven-cup wall machine. Bryan's Works continued manufacturing amusement machines until 1990. Although most of the machines built at Kegworth were ball and cup games, Bryan's were also responsible for the Live Peep-Show (1936), which used live ants as its attraction.

By 1935 there were nineteen major British slot-machine manufacturers (including makers of vending and weighing machines), but production peaked in the late 1930s and by 1951 the number of manufacturers of

Belt-driven machinery used in the manufacture of slot machines at the Bryan's Automatic Works of Kegworth, Leicestershire, between the 1930s and 1990. The machinery is now on display at the Bryan's Works Museum, Drayton Manor Park, Tamworth, Staffordshire.

Above: *The Flamingo arcade on the Marine Parade, Great Yarmouth, in 1991. Modern amusement arcades have such spectacular and expensive lighting displays that the seaside close season has disappeared and arcades open all the year round to maximise their income.*

amusement or vending machines had been reduced to six. Only 25 amusement arcades existed in London in 1950, compared with over 170 in the early 1900s. During the 1950s wholly electrical games were introduced into amusement arcades and the popularity of mechanical games began to decline, eventually resulting in the sale of many of them for scrap in the late 1960s. A resurgence of interest in early amusement-machine design followed during the 1980s, and pre-1950s machines are now collector's items.

Left: *The interior of a modern amusement arcade on Marine Parade, Great Yarmouth. Today's machines are larger, louder and noisier than their ancestors. Only a small range of machines is available in most arcades, but video games in particular are complex and can provide several minutes of entertainment.*

A working model of 'The Spiritualist Room', complete with 1930s furniture. When a coin is inserted the table levitates, a ghastly head appears behind a chair and the hearthrug (and the dog sleeping on it) rotate. The model was probably made in the late 1930s and used in the massive Galaland arcade at Scarborough.

FOR AMUSEMENT ONLY

The words *For Amusement Only* often formed part of the decorative features of late nineteenth-century amusement machines which offered the player various forms of entertainment or a cheap gift in return for a penny (sometimes a halfpenny) in the slot. These non-gambling amusements filled the early amusement arcades and comprised five main groups: visual amusements, fortune-tellers, physical amusements, crane skill games and sporting games.

An optical effect in the form of the camera obscura had been used as seaside entertainment since the early nineteenth century, but the first slot-machine sales of visual amusement were achieved by means of viewers, either mutoscopes which imitated moving pictures, or static picture viewers, often stereoscopes. Static viewers were manufactured well into the twentieth century; the Bollands' Art Deco style Auto

Stereoscope featured a series of 'Happy Moments', typical of the mildly erotic 'living pictures' presented by the viewers.

Inserting a coin into the mutoscope and turning a handle caused a reel of minutely differing photographs to rotate, flipping over rapidly before the eye of the player and providing the illusion of movement. The reel of photographs was encased in a wooden or metal drum resting on a stand; lighting was either by natural light or by electricity generated by turning the handle. Decoration on cast-iron models could be highly ornate. Arcade operators found that gently naughty or mildly exciting images were more profitable than educative films. The most famous title was 'What the Butler Saw'; others included 'Stolen Locomotive', 'I'll Say She Can' and 'The Naked Truth'. Despite the basic innocence of the contents, mutoscope operators often came into con-

Concentration on a mutoscope showing a boxing match in an arcade on the Palace Pier, Brighton, in the 1950s. Also available on these imitation films were 'Parisien Can Can' and 'All Alone'.

The internal workings of a mutoscope. As the reel of photographs was turned, normally by an external handle, the viewer saw each for an instant before it flipped past the eyepiece. These fleeting glimpses simulated movement, and a story could be told within the time a reel of photographs took to rotate.

flict with the authorities; a successful campaign was waged against indecent mutoscopes in Blackpool during 1899.

Working models, as the name suggests, provided the player with a brief animated play, often on an horrific theme, after the insertion of a coin. 'The Haunted Churchyard' (graves opened, skeletons and ghosts appeared) and 'American Execution' (featuring an electric chair) were typical titles. The models, contained in glass cases on stands, were rarely identical and often included finely detailed miniature figures and buildings. One of the best-known manufacturers was John Dennison of Leeds, who began making working models in the late 1870s and had the amusement-machine concession at Blackpool Tower from its opening in 1894 until his death in 1924; his daughters Alice and Evelyn continued the concession until 1944. They produced the carefully crafted and extremely unpleasant 'Chinese Torture' in 1938.

Nelson Abyssinia Theodore Lee produced working models at his home and workshop in Louise Street, Blackpool, from the late 1890s until about 1927. Lee (1869-1943) constructed the characters, moving parts and cases for his models, dressed the figures with the help of his wife and bought only the clockwork motors. His models are notable for their careful detailing and were on display in Lee's amusement arcade on the South Promenade at Blackpool until the 1930s.

The raffish atmosphere of the Palace Pier at Brighton, opened in 1899, partly resulted from its collection of working models, mainly produced during the 1900s by Nelson Lee's son Leonard Lee. His final and most popular work was 'The Guillotine'. More frequently seen than Lee or Dennison models are Bolland machines made during and after the Second World War, based on the motors from crane skill games which were then unused as prizes were unobtainable. Their titles included 'Haunted House' and 'Egyptian Tomb', but most manufacturers produced versions of the more lucrative titles like 'Drunkard's Dream' and 'Spanish Inquisition'.

Fortune-tellers, machines providing a horoscope for the player on receipt of a coin

Left: *The Oracle, an electronic fortune-teller.* Early horoscope machines often provided a printed card which purported to tell the player's fortune, but the advent of electronics introduced 'science' into the process. The player consulted the Oracle by turning the central knob to select the question, inserted a coin and turned a small handle below the respective coin slot. Internal lights then illuminated the answers one by one, coming to rest at the 'correct' answer.

Right: 'Shocking Jack', the electric sailor, a cast-iron early twentieth-century model probably made in Germany. A coin in the slot above his white breast pocket would cause Jack's eyes to glow and a small electric current to flow through the player who grasped his hinged arms in both hands.

in the slot, were amongst the most popular of the early amusement machines. The first patent for a coin-freed automatic fortune-teller was taken out in 1889, and contemporary ideas included audible fortune-tellers, clockwork figures issuing horoscope tickets, and a delightful 1891 suggestion that fortunes should be displayed on imitation bubbles floating in water inside a case made in the shape of a bar of soap. Most 1930s and 1950s fortune-tellers used automated gypsy figures in mock fairground booths which delivered printed horoscopes.

Some physical amusements used the novelty of electric shocks to entertain the player, but most involved the player lifting, pulling or hitting parts of the machine, the score being shown on a dial incorporated into the decorative casing. Bulls and owls were amongst the themes used for strength-testers, and punch balls were widely manufactured from 1904.

Crane skill games, in which the player attempted to obtain a small prize by manipulating a crane and grab via an external control, were invented in the United States in 1924. The game was contained in a freestanding casing, often Art Deco styled, with glass upper sections through which the player could see the crane hanging over a display of small gifts or sweets. Devious operators might include high-value prizes in the display but place these out of reach of the grab or even glue them down, thus encouraging repeat plays. An important manufacturer of crane skill games in the 1930s was the Exhibit Supply Company of Chicago.

Left: *A counter-top grip-testing machine. The single-handed grip test took up less arcade space than punch balls and other freestanding strength-testers. The player gripped the handle as hard as possible, causing the pointer to move and give a score. The average grip for certain professions was given on the two cards below, ranging from a banker (weak) to a sailor (strong).*

Right: *A modern reproduction of the Mills Perfect Muscle Developer or Owl Lifter, first manufactured in 1904. Insertion of a coin freed a locking device on the pair of handles attached to a platform at the base of the machine. The player pulled upwards on the handles to record a score, between 'Not so good' and 'Great stuff, big boy!', and the handles locked again on release.*

Mechanical music machines were the forerunners of the juke-box. During the late nineteenth century they were often found in public houses and some featured a working model, often a singing bird. The addition of a mechanical music function was used in America as a successful ruse to disguise gambling machines from the authorities during times of near prohibition. Manufacturers attached plates to the machines stating that they were intended only for playing music and that any money won incidentally had to be reused in the machine. Conflict with the law was a constant problem for amusement arcades in the early twentieth century, with court cases turning on whether machines provided games of skill or chance. Application of the law was far from consist-ent, though it could be swingeing: in May 1912 in Sunderland 121 men, women and children were arrested in an arcade, 109 of whom were later charged with illegal gambling.

The earliest sporting games were races, the players betting amongst themselves on the outcome rather than receiving any payment directly from the machine. William L. Oliver's patent of 1887 was the first of this type of mechanism and featured two or more horses powered by clockwork, weights, electricity or the players themselves, racing in concentric tracks; the players placed bets on the outcome. In 1893 race games were adapted for the single player by means of a machine payout; the player had to propel his own horse faster

A page from the January 1946 Bryan's Works of Kegworth catalogue showing three of their range of crane skill and similar games: (from left) the Six-Sided Crane, the All-Square Crane and the Prize-Finder. The catalogue was the first to be issued by Bryan's following the Second World War, during which time the factory was turned over to munitions production, while W. E. Bryan himself served in the RAF.

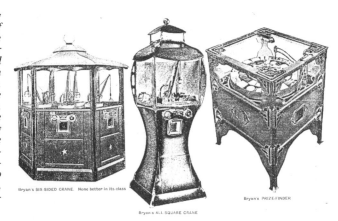

Bryan's SIX-SIDED CRANE. None better in its class

Bryan's ALL-SQUARE CRANE

Bryan's PRIZE-FINDER

Below left: *The Polyphon, a mechanical music machine which used a metal disc connected to a musical box to produce a tune. The holes in the disc corresponded to musical notes. This model was made in Leipzig, Germany. Polyphons were often to be found in the bars of public houses in the 1890s.*

Below right: *The Clucking Hen which delivered a coloured metal egg containing a small toy or charm on receipt of a coin. This German machine was popular in England during the 1920s and 1930s, when over five million eggs were produced each year.*

NEW LAID EGGS
PUT 6ᵈ IN SLOT
PULL HANDLE AND YOU WILL GET
A SURPRISE EVERY TIME
LISTEN TO THE HEN CACKLING.

Above left: Incongruous classical decoration on the metal casing of a Penny Shooter, a miniaturised version of a shooting range. Some shooting games used real projectiles, either coins (popular in the 1890s) or balls, while others registered 'hits' via a mechanical link. The idea of the shooting game has now been taken up in complex video games.

Above right: The two competing cyclists in Marathon Cyclist (also known as the Grand Marathon Cycle Race) made by Charles Ahrens of London during the 1930s. Two players turned handles to power their cyclists around the vertical circular track, and the winner had his or her stake returned. The centre of the track was filled by a cycling illustration, sometimes a painting rather than a mass-produced image.

Left: Part of the 1901 patent specification by J. G. M. Pessers for a shooting game in which the player had to direct the soldier's bayonet into the eye of the oriental target. The coin-freed mechanism is connected to a spring which causes the soldier to move up and down for a predetermined length of time. The trigger (T) releases the figure to plunge towards the target, and a direct hit on the eye results in a prize. The 1900 Boxer Rising in China, crushed by Britain and her allies, probably explains the symbolism.

than a clockwork horse to win. The first race games featured horses, but later dogs, bicycles, athletes, yachts and by the 1930s cars were introduced.

The popularity of the live shooting gallery continued into the twentieth century but slowly waned as mechanical shooting games, lacking the inherent dangers of live ammunition, became available from the 1890s. The first shooting games involved firing balls, pellets or other objects at targets; later shooters were electronically controlled, and no actual projectile was fired.

Contests between players or teams not involving races or shooting could be mim-icked using air pressure or direct manual operation. E. G. Matthewson was the pioneer in this field with a football game patented in 1896, followed in 1899 by a cricket game, but both used only two model players. It was 1921 before George and Harry Barr with the British American Novelty Company produced a football game using full eleven-a-side teams. These machines used a basic single kicking-leg mechanism for the model players and had large wooden cases, often with cast-iron claw feet. The models wore tiny knitted jumpers in team colours.

Left: Full Team Football, a competitive game for two players manufactured by the British American Novelty Company of London in 1921. This was the first of many freestanding competitive games with sports themes which were popular throughout the 1920s and 1930s. The sports imitated included polo, boxing, greyhound racing and many more; one novel contest was racing to fill a pint glass with 'beer' in The Bar, made by Walton & Company of Blackpool in 1929.

Right: Bryan's Worl Borl, a competitive football game for two players first manufactured in 1953. It used seven small metal balls, around half the diameter of typical wall-machine balls, which the players drove around the playfield one at a time by means of bats positioned to simulate the kicking action of the tiny footballers. The bats were controlled by knobs at the bottom left and right of the casing. A goal was scored if the ball passed behind the opposition footballer into the area beneath his feet; the game was for amusement only.

Left: *The man with the keys was a constant presence in the arcades of the 1950s and 1960s. Wall machines were prone to occasional malfunctions, and all seemed to need a different key. Young players found it worthwhile to check the prize-delivery trays of machines when not in use, as surprisingly players often left their winnings behind.*

Right: *A detail from the American counter-top game Spiral Golf. A trigger action fired the metal balls round the spiral playfield, the balls scoring when passing through any holes but the inner and outermost, the 'ruff' and lake. The player had five balls to make a par of 500, when each ball scored in the hole next to the central hazard. Rather than the machine paying out, winners claimed their prizes from over the shop counter.*

WALL MACHINES

Amusement machines which were attached to the walls of amusement arcades, as opposed to being freestanding, were known as wall machines. They normally consisted of a wooden case containing the mechanism fronted by a vertical glass panel revealing the outer workings of the machine to the player. The playfield was slightly slanted away from the player to avoid the need for extra lighting within the machine case. There were also a coin slot, often at the top right-hand side of the case, a delivery tray for prizes at the bottom of the machine, and various combinations of sprung trigger, ball-release handle and controlling levers,

depending on the details of the design. Early models gave a free play for a win, and later small prizes in cash or kind were introduced. Cigarettes were frequently used as prizes from the mid 1920s, and sweets were popular from the mid 1950s to the late 1960s. The great majority of wall machines used a metal ball of roughly $1/2$ inch (12 mm) diameter as the projectile, the player being required to direct one or more balls into cups, slots or other winning destinations.

In the United States, where wall machines were never popular, the counter-top game was the equivalent. Counter-top games had similar mechanisms to wall machines but on a smaller scale, their cases usually being around 18 inches (457 mm) high, with winning payments made over the shop or bar counter rather than through the machine.

Wall machines can be divided into four basic categories: drop-case machines which used the coin as a projectile; variations on Tivoli billiards with skill pins to deflect the ball or coin; those where the player could move the winning receptacle; and the simple ball and fixed cup type. By the 1950s wall-machine design had evolved to include internal operators' skill controls. These varied the amount of win payments (combinations of coin return, ball return and cash or prize), the size of the jackpot and the ten-sion in the trigger spring; there were also anti-cheat devices to guard against rough handling of the machines.

Drop-case machines, which used the coin as a projectile and thus had relatively simple mechanisms, were patented from 1892 onwards in a wide range of designs. The coin was normally inserted at or near the top of the machine and fell towards the winning area past horizontal skill pins or other barriers. Elegant background designs were a feature of these machines; Fred Bolland's 1920s model The Lifeline showed a lighthouse and seascape complete with seagulls. Drop-case machines were the precursors of the 1950s Pushers, pioneered by Cromptons of Ramsgate, in which coins dropped down

Left: *Sky Jump, a much-altered wall machine probably dating from the 1950s and working on the drop-case principle, in which the coin-released metal ball dropped down the centre of the playfield rather than being fired by the player. Turning the knob on the right of the case changed the angle of the barriers as the ball dropped, the object being to direct the ball away from the centre and side losing pockets.*

Right: *A wall machine giving full value for one penny, as the player was given six attempts to manoeuvre a falling ball into the centre column, using the movable paddles worked by external handles. If the player was successful every time, the sixth ball fell through a hole and triggered a prize or replay mechanism.*

 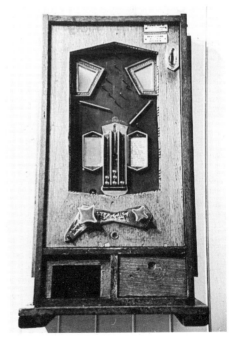

Left: *The Clown, a well-known ball and moving-cup machine in which the player tried to catch a ball in the clown's hat. The player could move the clown horizontally while the ball fell through the skill pins. This version originated in Leipzig but was reconditioned by Clem's Automatics of Redcar after the Second World War. They altered the mechanism to provide a cigarette prize.*

Below left: *B. W. Brenner's unusual wall machine Get the Ball Past the Arrow, popular in arcades during the late 1950s and 1960s. It is a modernised version of Tivoli billiards with a striking black wooden case and a background design featuring red and yellow shooting stars. The arrow moved to give a different win area on each play.*

Below right: *The interior of Get the Ball Past the Arrow, showing the machine mechanism attached to the door of the machine, the inside of the playfield. The coin entered at top left, releasing the central springs and levers controlling the arrow. The trigger, bottom left, fired the ball around the playfield, and coins travelled down the vertical chute, lower centre, to the winning player.*

by players from several playing stations were diverted by pins and piled up on a moving surface below; any addition to the pile could cause it to topple into the winning chute.

Variations on the old English game of bagatelle were often named according to the arrangement of their hazards of pins, holes and hoops; during the late nineteenth century bagatelle in a small vertical format came to be known as Tivoli billiards. This game was ideal for conversion as a wall machine, and the pioneer inventor Henry John Gerard Pessers produced the first patented version in 1899. The player fired a series of balls upwards in a channel at the side of the machine; the balls then fell through an arrangement of retaining pins, a winning combination completing an electric circuit to give a prize. Numerous variations were quickly patented; one by Haydon and Urry's Automatic Machine Company followed in 1900 and then several more each year until by 1914 eighteen separate designs were registered. This represented the peak of interest in Tivoli billiards machines, which were gradually replaced in the arcades by ball and cup machines during the 1930s.

Machines with a movable winning receptacle had more complex mechanisms than ball and fixed cup designs and were re-

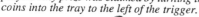

Left: *Elements of Art Deco and classical design combine in the playfield and casing of Playball, an Allwin-type machine with a horizontal line of seven winning and losing cups. In this very basic form of wall machine the ball went out of play only via the losing cups; more complex designs incorporated a losing hole at the bottom of one of the spiral ball guides, thus a ball missing the line of cups was normally lost. In Playball, if the ball misses the cups it is simply returned to the player.*

Right: *The Bryan's Automatic Works machine Fivewin ready for action, showing the metal ball about to be triggered around the playfield. With deflecting pins positioned above the cups and only five winning cups out of seven, the odds in this Allwin-type machine were not in favour of the player. Any prize was claimed by turning the knob at the lower left of the casing, which released coins into the tray to the left of the trigger.*

Bryan's Works of Kegworth wall machines. Fivewin (top left), Elevenses (bottom left), Pilwin (top centre) and Ten-Cup (bottom centre) were shown in their 1946 catalogue and were variations on the same theme of a ball dropping into winning cups. As many other manufacturers were producing similar designs at the time, Bryan's emphasised the durability of their machines as one of their prime selling points. W. E. Bryan rang the changes on wall-machine design with his Gapwin (top right) and U-Win (bottom right), which both had large win areas rather than sets of small win cups. The catch for the player in Gapwin was that two levers initially covered the win area; the player had three balls to move the levers and score. U-Win apparently gave a large chance of winning, but the machine was designed so that a ball driven round at normal speed rarely dropped near the win area; skilled players manipulated the trigger in order to slow the ball down.

garded as games of skill; in pure ball and cup games chance rather than skill was the deciding factor, though arcade operators argued otherwise. In a movable-cup machine the ball was normally fired upwards by the player, who then tried to anticipate its route as it fell through a series of obstacles, usually skill pins, and catch it in the movable cup before it fell into the losing area below. The Pickwick (from 'pick quick') was the first of these machines, patented by H. J. G. Pessers in 1900, but best known is The Clown, introduced in 1905 by Max Jentz and Meerz of Leipzig, Germany. The clown's upturned hat acted as the receptacle for the ball.

The Pickwick was at the centre of a 1911 court case in which the prosecution alleged that the machine was a game of chance rather than skill. The machine in question was installed in the bar of the Spon Lane Tavern, Birmingham, and successful players received metal checks worth 2d which could be exchanged over the counter for beer or tobacco. Groups of players also used the machine together, placing winnings and forfeits into a kitty which was later exchanged for beer; at times, more money was taken through the machine than was paid over the counter by customers. The case ended with the brewery agreeing to remove the machine from the public house, although the Pickwick was not proved to be a game of chance. Indeed, the following year the Pickwick was adjudged in court to be a game of skill, although this did not prevent further prosecutions against amusement machines being attempted.

Ball and cup machines form the largest group of wall machines, manufacturers hav-

Right: *The Trickler, shown in the Bryan's of Kegworth 1946 catalogue, was patented in 1934. Ten balls dropped singly from the top of the machine and the player attempted to catch them in movable numbered cups. Although it was intended for the single player, it was advertised as a game for clubs and pubs, 'or anywhere where a few of the fellows gather together'. Players would bet against each other to see who could achieve the highest score.*

Below left: *A seven-cup Allwin, probably produced in the late 1950s or 1960s by Oliver Whales of Fun City, Redcar, Cleveland. The prize was a small chocolate Aero bar, and the playfield background uses an aeroplane design as a pun on the prize.*

Below right: *A basic Allwin with seven winning cups, probably produced by Oliver Whales at Redcar in the late 1950s or 1960s. The prize was a KitKat chocolate bar, which the winning player obtained by turning a knob on the left of the wooden casing. Although the casing design was strictly functional, the playfield was colourful and included an elegant tree-shaped Rowntree symbol.*

19

Bryan's of Kegworth produced this unusual wall machine known as the Gapwin, which had a large central winning cup containing three separate winning holes. The player had three balls to move the pivoted bars obstructing the gap and send all three balls into the winning holes. The bars were not reset for each new game, thus encouraging the player to invest in further attempts if not at first successful. Balls disappeared from play only through the two large losing cups.

ing produced an enormous number of design variations on the simple theme of balls dropping into cups. Although Haydon and Urry registered a patent for a ball and cup design in October 1900, it was not until the 1930s that the manufacture of these machines, cheap and easy to make, became widespread. Popular ball and cup designs were extensively copied and old models adapted for new prize payouts. These alterations, combined with company mergers during the 1930s and the production of machines by individual amusement arcades, make identification of wall-machine manufacturers difficult; often the only name on the machine is that of the supplier.

One of the simplest and most copied ball

and cup designs was the Allwin, which originated in Leipzig. Insertion of a coin freed a ball which the player sent round a spiralling metal track with a sprung trigger; the ball eventually lost momentum and fell on to a line of winning and losing cups. An Allwin can be strictly defined as a ball and cup machine using a single line of cups, although machines with other arrangements of cups are sometimes known by the same name. English versions include the Allwin de Luxe, produced by both Frank Harwood of Birmingham and Oliver Whales of Fun City, Redcar, Cleveland. These Allwins used five winning and two losing cups, arranged in a line with the losing cups at either end.

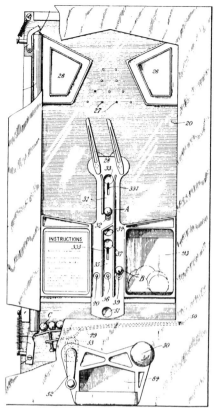
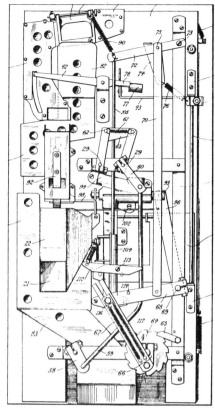

Left: *The playfield patent drawing of William Bryan's Payramid, invented in 1934 and patented the following year. It became the best-known of all the variations on collecting balls dropping through pins from the top of the machine. The gap between the movable fingers, pivoted by turning the knob on the right, could be adjusted to change the degree of difficulty. This version offered a jackpot if all eight balls were caught.*

Right: *The mechanism of the Payramid, one of thirteen diagrams necessary to explain its workings in the patent specification.*

HOW SLOT MACHINES WORK

The basic principle of the slot machine is that a coin inserted by the player into a slot in the machine casing frees the main part of the mechanism to function. Primitive coin-freed vending machines first appeared in Britain in the 1830s, but it was not until 1857 that Simeon Denham of Wakefield applied for protection by patent for a coin-freed mechanism, designed to sell strips of stamps. Automated vending became popular by the late 1880s, as did attempts to defraud machines by tampering with the mechanism or substituting coins of lesser value, buttons and washers.

To combat this problem, coin chutes were designed with holes, so that undersized coins would fall through without triggering the mechanism, while the use of heavy coins or other objects was defeated by means of a balance, set to be tripped by overweight coins; a coin of correct weight would continue on its way into the machine. Cheating by tying the coin to a string was deterred by incorporating zigzag coin chutes and string cutters into the mechanism. Successful use of a bent wire as the trigger

Left: *Two machines which Bryan's produced for many years, the Clock (left) and the Payramid, shown in their 1946 catalogue. Although machine casings were altered to keep up with fashion and mechanisms were slowly improved, the playfield of a popular machine generally remained unchanged. A visible jackpot, for example the coins on the lower right of the Payramid playfield, attracted potential players.*

Below left: *The mechanism of a Bryan's Automatic Works Twelvewin Clock. To the right of the clock face are the bellows used in the original 1930 design of this machine to slow down and stop the movement of the clock hands.*

Below right: *A later version of the Bryan's Twelvewin Clock mechanism, in which the unreliable bellows have been replaced by a system of springs and levers. The Clock was produced in various forms and a variety of casings until at least 1960 at the Kegworth works.*

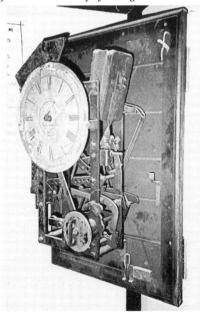
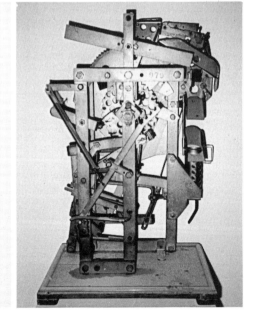

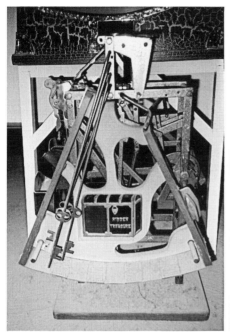 

Left: *The mechanism of a Bryan's Hidden Treasure machine, first produced in 1938. When a coin was inserted the four pendulums swung to and fro for a short period, and the player won if one of the keys came to a stop exactly over a keyhole on the casing. Most Bryan's machine mechanisms were constructed of steel, gunmetal and tinplate and were purely mechanical (rather than electronic), thus reducing the possibilities of malfunction.*

Right: *The interior of a Copper Sega fruit machine, showing the three reels and the complex system of interconnecting springs and levers. Sega Incorporated of Japan was founded in 1957, and its first machine was a copy of an American Mills Hightop.*

could be avoided by designing a two-part mechanism which only a coin could connect. Players also used force, simply hitting the machines, blew down the coin slots to trigger the mechanism, and used electromagnets to bring the mechanism to a halt. The early years of vending and automatic gaming machines were a constant battle between owners and potential cheats.

Despite the many attempts to deter fraud, the efficient and reliable sale of high-cost items by vending machine never became a real possibility, and by the 1890s coin-freed mechanisms were in use only for the sale of cheap goods and in gambling machines. Although an American patent of 1876 made the first suggestion of a coin-operated gam-bling game, William Oliver's 1887 horse-race patent was the first serious application of the coin-freed principle to a game of chance. Automatic win payouts were first introduced on American machines in 1889; the earliest British example was Frank Urry's 1892 Tivoli billiards design using the coin as projectile. It automatically provided the winning player with a cigar or similar reward. The mechanism for delivery of the prize entailed the coin falling on to a lever which freed a pivoted column; the player then pushed on a handle, which forced the column backwards and ejected the prize into a tray, while the coin fell into the cash till and the lever was reset.

Amusement-machine mechanisms were

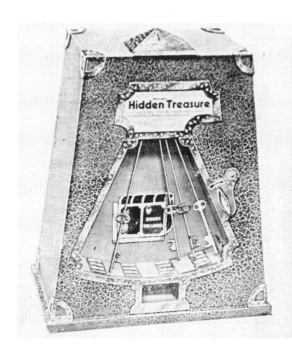

almost as varied as playfield designs and were fine small-scale examples of mechanical engineering. The ubiquitous wall machines of the 1930s and 1950s often used small metal balls released by a coin-freed mechanism; the coin inserted by the player rocked a lever controlling a ratchet wheel, allowing the wheel to turn and a ball to escape from the storage chute into play. This mechanism, patented by Rudolf Walther in 1913, formed part of many later Allwin machines.

Superficially, the Allwin appeared an attractive proposition, with several win cups and few or no losing cups, but skill pins sited just above the joints between the cups often deflected the ball into the losing area below the line of cups. The ball had to drop almost vertically into the cup to be retained; at any other angle, the ball tended to bounce off the sides of the cup. An Allwin de Luxe, with five winning and two losing cups, had apparent odds in favour of the player of $2^1/2$:1, but the real odds were at least 3:1 in favour of the machine, taking the layout of the playfield into account.

The Bryan's of Kegworth Four-Square, showing Pilwin (left) and Gapwin. The bright colours of the playfields and the attractive finish of the whole assembly made the Four-Square a good proposition for an arcade owner.

 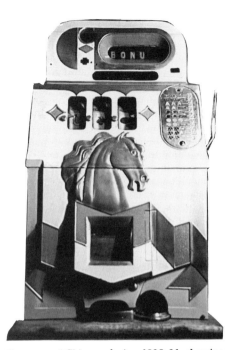

Left: *The Poinsettia, made by the Mills Novelty Company of Chicago during 1928-31, showing the flower decoration on the metal casing which makes this model a favourite of collectors.*

Right: *The Bonus, an unusual gambling machine produced by Mills in 1937. It paid out an extra prize if the word BONUS was completed in the top window. The casing decoration combined classical imagery and Art Deco styling, tradition and modernity. The machine was still being advertised for arcade use in the 1950s.*

GAMBLING MACHINES

Early gambling machines in Europe and America were usually variations on the spinning dial and were known as roulettes. The dial was marked with numbers, colours or symbols and began spinning on insertion of a coin; a pointer indicated the winning segment when the wheel came to a halt. More complex machines had two or more concentric dials, particular combinations of symbols giving payments. A British patent for a mechanism with multiple dials and a viewing window for the final combination of symbols was registered in 1890 and was a clear precursor of the fruit machine.

Roulettes could be freestanding, wall machines or counter-top models. The Mills Big Six, first made in 1906 by the Mills Novelty Company of Chicago, was a tall, freestanding and beautifully decorated machine in baroque style with claw feet

topped by winged lion's heads. Caille Brothers of Detroit were one of the most important early makers of roulettes, which are now often known as Cailles or Caillies irrespective of their actual manufacturer. Caille's counter-top roulettes were widely copied in Britain in the 1930s as wall machines.

Although manufacture of roulettes continued after the 1930s, they had already been superseded in popularity by the spinning reels of the fruit machine. Invention of the multi-reel slot machine with automatic payout is generally ascribed to Charles Fey of San Francisco, who developed it during the period 1895-1905. His original machine, the Liberty Bell, was copied with some alterations and mass-produced by H. S. Mills of Chicago, who set up the Mills Novelty Company. Mills had begun making amusement machines in 1889, and by 1932 his company was manufacturing 70,000 machines a year.

The typical fruit-machine mechanism consisted of three reels, on the edge of which were a variety of symbols, often stars, fruit or bells. Each reel had a timing device, actuated by the insertion of a coin; when the handle of the machine, normally located on the right of the casing, was pulled the reels spun and were stopped after an interval by the timer. The symbols were visible through a window in the casing. Winning combinations of symbols were shown on a card attached to the casing, and the payout was automatic, by means of perforated metal plates attached to each reel; a win caused metal fingers to project through the holes, tripping a coin slide. Later alterations included the introduction of five-reel machines, jackpot payouts and electric machines with push-button starting. The term 'one-armed bandit' arose from the original design of the machine and the ability of operators to adjust the rate of win payments downwards; unscrupulous operators could also fix a 'bug' (so named

from its shape) to a reel at a selected point to prevent the reel stopping where a jackpot symbol appeared.

It was 1926 before the first fruit-machine patent was filed in Britain, but by the early 1930s registration of American fruit-machine designs in Britain was common-place. Apart from the Mills Novelty Company, important American makers were the Watling Manufacturing Company of Chicago, Caille Brothers of Detroit, O. D. Jennings (a former Mills employee), the Pace Manufacturing Company and the Bally Manufacturing Company of Chicago, whose electric fruit machines now dominate the American market.

The particular delight of fruit machines lies in the design of their metal casings. The Mills Liberty Bell was covered in Art Nouveau decoration and the 1935 Watling Rol-A-Top had a cascading coin motif which endeared it to collectors. Mills also produced the Poinsettia in 1928-31, with a casing wreathed in flowers and leaves. Art Deco styling was popular from the 1930s, both the Mills Extraordinary and the Copper Sega having streamlined styling.

Left: *The Film Stars, a single-reel fruit machine using the names of film stars like John Wayne and Betty Grable as winning symbols on the reel; Ingrid Bergman gives the top prize. The unusual wooden case has stark Art Deco styling.*

Right: *The interior of the Film Stars, showing the relatively simple mechanism of a single-reel fruit machine. One in five of the names on the reel is a winner, but as the highest win (which appears only once on the reel) gives six times the stake, the odds are at least 2:1 in favour of the machine.*

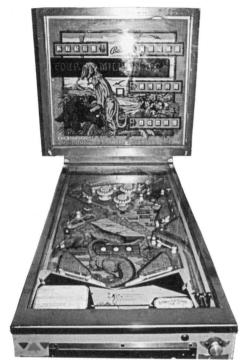

Left: *Bally's Four Million BC, first produced in 1971 and designed by Ted Zale, Bally's sole pinball designer from 1963. The artwork, on a prehistoric theme, was outstanding and the game featured Zale's 1966 invention 'zipper flippers', a pair of flippers which closed up when the correct target was hit. On the right is the backflash of Magic City, showing an elegant fountain in a street of skyscrapers.*

Below left: *The playfield of Four Million BC. The ball emerges from the plunger lane on the right and crosses the middle of the playfield, rather than being sent round the top of the playfield as in most designs. Scoring features included the volcano (middle left), the tar pit (top right) and three bumpers (top centre) which repelled the ball.*

Below right: *The interior of Four Million BC, showing the complex electronic controls beneath the playfield. The entire playfield lifts towards the backflash for access.*

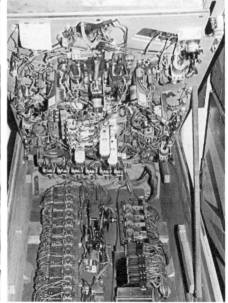

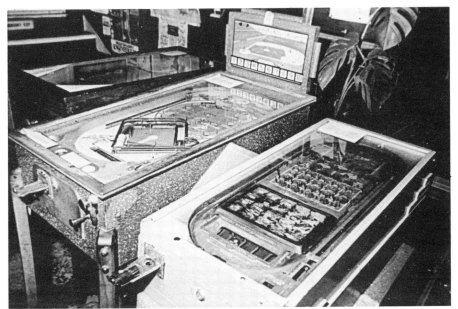

Two pinball machines from the 1930s. On the left is an electromechanical game dating from 1934-5 which uses images of baseball; the backflash shows a baseball diamond as well as the score. The early games World Series and Fifth Inning (1939) shared the baseball theme. On the right is a version of the mechanical Jigsaw game first produced by the Rock-ola Manufacturing Corporation in 1933 to celebrate the Chicago Exposition of 1933-4. The jigsaw was completed by the action of balls which flipped over individual pieces after falling through the correct holes.

PINBALL MACHINES

Between the 1930s and the 1960s no British amusement arcade was complete without its pintables or pinball machines. Pinball, a development of bagatelle, became a mass-market success in the United States from December 1931, when David Gottlieb, a Chicago slot-machine manufacturer, introduced a bagatelle variant called Baffle Ball. Coin-operated counter-top bagatelle games in vertical and flat playfield forms had been marketed since around 1900, but by combining a low selling price with a design attractive to both operators and players Gottlieb took bagatelle into a different market.

Baffle Ball was a purely mechanical game, with an arrangement of pins and holes on the slightly sloping playfield which enabled players to amass a score with their seven balls; it cost one cent per game. Other slot-machine manufacturers immediately pro-

duced their own versions of bagatelle, and the game quickly developed to include first mechanical and then electromechanical hazards and scoring opportunities. The backflash, initially merely a means of showing the score, became by the late 1930s a brightly lit advertisement for the game. The size and slope of the playfield became standardised, at 19¼ inches (489 mm) wide by 36¼ inches (921 mm) long, with an incline of 3.5 degrees towards the player. It also became ever more colourful, and the game itself, played with steel balls 1¹/₁₆ inches (27 mm) in diameter, grew more complex. Tilt controls, which guarded against the player cheating by nudging the machine too violently (using 'body English'), were introduced in 1934.

The first advertisement for pinball machines in Britain appeared in the showmen's paper *World's Fair* on 23rd July

Left: *The backflash of Buckaroo, a Gottlieb machine first produced in 1965. The artwork is based on a Wild West theme, and high-scoring players could win a replay; other games gave only extra balls as a prize.*

Below left: *The Buckaroo playfield with (centre) its Spin Roto, a revolving target which varied the score gained when hit. Spinning targets could be horizontal or vertical in relation to the playfield, and some had to be activated by hitting another target. Other types of target could drop beneath the playfield or swing from side to side.*

Below right: *The reverse of the Buckaroo playfield showing the vertical spinning target projecting downwards into the machine's interior.*

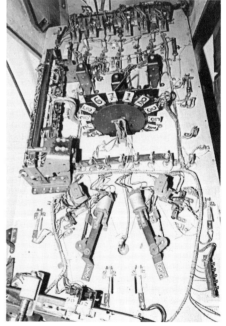

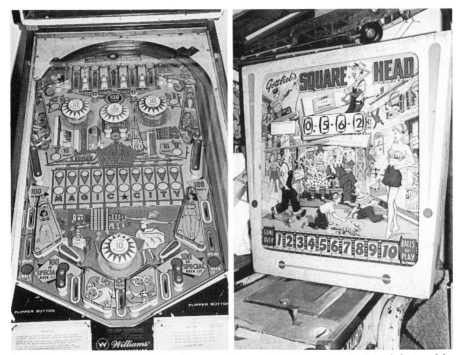

Left: *The playfield of Magic City, first manufactured by Williams in 1967 and designed by Norman Clark. The sophisticated urban theme of the artwork included a chic restaurant, American footballers, skyscrapers and a nightclub. Pinball-machine designers were generally responsible for the working of the playfield and the rules of the game, but not for the artwork.*

Right: *The backflash of Gottlieb's Square Head, designed by Wayne Neyens in 1963. The artwork theme was a cartoon-style street scene which included the current score and number of balls left in play. The playfield featured nine bumpers, making for constant action as the ball was buffeted around.*

1932. The introduction of the game to Britain revitalised the whole of the country's slot-machine market, and British manufacturers such as Burrows Automatics and Hardinges attempted to produce their own pinball machines, some little more than copies of American designs, in the early years of the craze. The British contribution to pinball design and manufacturing ended with the coming of war in 1939, leaving the Chicago manufacturers to achieve world market domination in the 1950s.

Post-war machines became more intricate. Flippers, small button-controlled bats enabling the player to send the ball back up the playfield, were introduced in 1947. Players enjoyed the flipper-equipped games and newly animated playfields; with a multitude of game designs available, the 1950s was the golden age of pinball.

The decline of the pinball machine began in the mid 1970s, when pinball was overtaken in popularity by the computer-controlled video game. Pinball had almost disappeared from British arcades by the late 1980s. Modern pinball machines incorporate a multi-level playfield, animated backflash and voice reproduction as well as assorted noises to accompany play; this new wave of machines is making a comeback in pubs, but few are to be seen in the arcades yet.

FURTHER READING

Baudot, J. -C. *Arcadia: Slot Machines of Europe and America.* D. J. Costello, 1988.
Colmer, M. *Pinball: An Illustrated History.* Pierrot Publishing, 1976.
Costa, N. *Automatic Pleasures: The History of the Coin Machine.* Kevin Francis, 1988.
Flower, G., and Kurtz, B. *Pinball: The Lure of the Silver Ball.* Apple Publishing, 1988.
Gresham, J. *Penny Arcadia Museum Guide.* Penny Arcadia, 1985.
Lindley, K. *Seaside Architecture.* Hugh Evelyn, 1973.
McKeown, H. *Pinball Portfolio.* New English Library, 1976.
Natkin, B. C., and Kirk, S. *All About Pinball.* Grosset & Dunlap, 1977.

PLACES TO VISIT

Museums and other venues with amusement machines permanently on display are listed below. However, displays may be altered and readers are advised to telephone before visiting to check that relevant items are on show, as well as to find out the times of opening.

Drayton Manor Park, Tamworth, Staffordshire B78 3TW. Telephone: 0827 287979. The Bryan's Works of Kegworth Museum shows the history of the penny slot machine and includes a reconstruction of part of Bryan's Works.

Eastbourne Heritage Centre, 2 Carlisle Road, Eastbourne, East Sussex. Telephone: 0323 411189. Small display of working slot machines.

Museum of the Moving Image, South Bank, Waterloo, London SE1 8XT. Telephone: 071-401 2636. Several mutoscopes, etc, mostly reconstructions, which visitors may play.

National Museum of Penny Slot Machines, Luton Fort, Magpie Hall Road, Chatham, Kent. Telephone: 0634 813969. A number of amusement machines.

National Museum of Penny Slot Machines, Newhaven Fort, Fort Road, Newhaven, East Sussex BN9 9DL. Telephone: 0273 517622. About thirty machines.

National Museum of Penny Slot Machines, Remember When, The Pier, Hastings, East Sussex. Telephone: 0424 422566. A number of amusement machines.

Old Penny Pier Arcade, Wookey Hole Caves, Wookey Hole, Wells, Somerset BA5 1BB. Telephone: 0749 72243. Extensive display of working amusement machines, many originally from Weston-super-Mare Pier.

Penny Arcadia, The Ritz Cinema, Market Place, Pocklington, York YO4 2AR. Telephone: 0759 303420. Extensive display of amusement machines from Jon Gresham's collection.

Woodspring Museum, Burlington Street, Weston-super-Mare, Avon BS23 1PR. Telephone: 0934 621028. Victorian Seaside Gallery with small collection of Edwardian slot machines, including mutoscope, working models and football game.

York Castle Museum, Tower Street, York, North Yorkshire YO1 1RY. Telephone: 0904 653611. Five working models on display (in use) including 'English Execution' and 'Spiritualist Room'.